La Vida Fantástica de Nacho

The Fantastic Life of Nacho

Written by the Spanish II students of South Holt High School.

Amazing Things Press

Book design by Julie L. Casey

Written and translated by Señora Rosemary Kerns
and the Spanish II students at South Holt High School, Oregon, Missouri:
Dalton Rauch
Madeline Prussman
Olivia Richards
Sophia Richards
Colleen Rukavina
Hayley Fansher

ISBN 978-1945667947

Printed in the United States of America.

For more information, visit

www.amazingthingspress.com

¡Hola! ¡Me llamo Nacho!

Hello! My name is Nacho!

Soy una iguana de Mexico.

I am an iguana from Mexico.

Señor Casey es mi "papá."

Mr. Casey is my "dad."

Tengo ocho años.

I am eight years old.

Tengo una gran vida.

I have a great life.

Recibo comida y attention.

I get food and attention.

Me gusta probar muchas cosas.

I like to taste a lot of things.

Soy un herbívoro. Me gusta la lechuga.

I am an herbivore. I like lettuce.

No como carne.

I don't eat meat.

La clase de ciencias es mi hogar.

The science class is my home.

Me gusta escalar en las cosas.

I like climbing on things.

Soy un amigo de los otros animales.

I am a friend to the other animals.

Mi mejor amigo es Pablo, un bearded dragon.

My best friend is Pablo, a bearded dragon.

Me gusta la luz caliente.

I like the warm light.

Se siente tan bien.

It feels so good.

Me da mucho sueño.

I get very sleepy.

Buenos noches.

Good night.

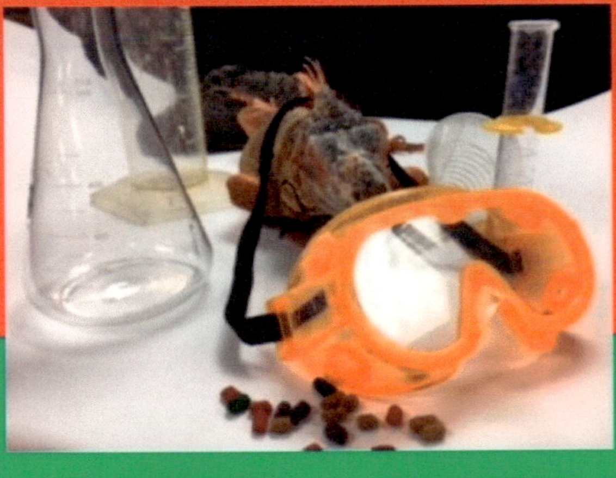

Descubre más sobre Nacho y sus amigos animales.

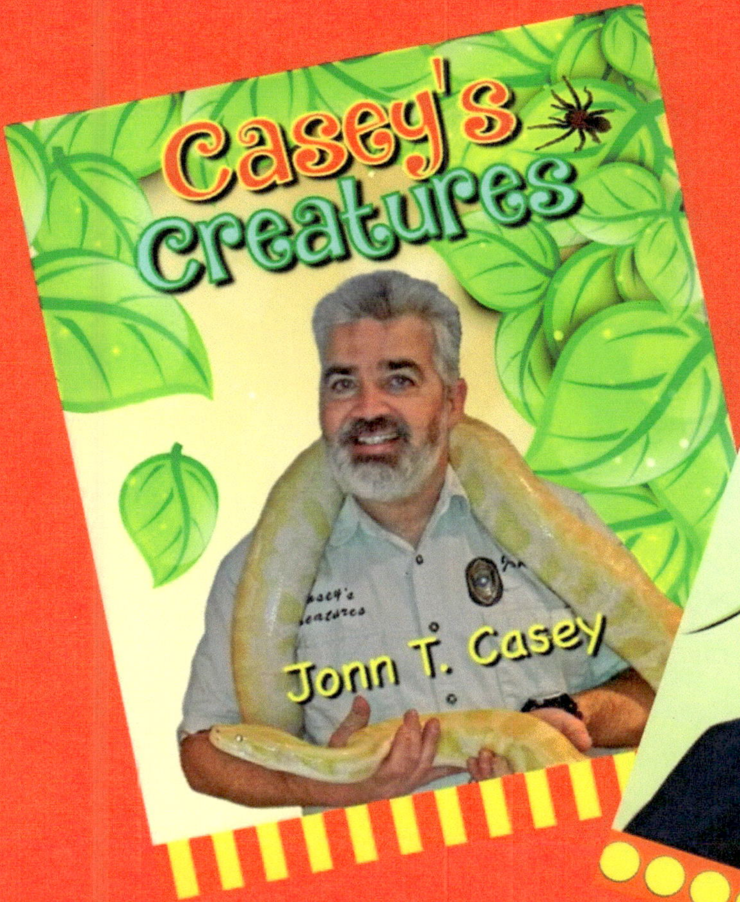

Find out more about Nacho and his animal friends.

Check out these Children's Books from

Amazing Things Press

Fancy Encuentra un Hogar by Kathy Hendricks

Catapillers by Kathi Boylan and Sharon Lea Larsen

The Neighborhood by Thomas Kirschner

Baby Noah by Sharon Lea Larsen

Sweet Potato by Kathi Boylan and Sharon Lea Larsen

Froggie Greenjeans by Kathi Boylan and Sharon Lea Larsen

Little Black Bear Cub by Rita Clark

Messy Bessie by T.R. Henson

Picky Betty Butbut by T.R. Henson

Geography Studies With Animal Buddies USA by Vashti Daise

Hoppity by Kathy Blair

Dougie the Dragon by Kathy Blair

Carter Symanski Cleans Under his Bed by T.R. Henson

Zany Zelda at Zitherfest by T.R. Henson

If I Had My Druthers by T.R. Henson

Mr. Wiggles' First Christmas by Shivonne Jean Hancock and Jane Neill-Hancock

Falling Water by Thomas Kirschner

Donald the Dog by Shivonne Jean Hancock and Jane Neill-Hancock

Penny the Parakeet and the Squadron of Squirrels by Melody L. Schaefer

Radiant Red by Vanessa Culver

Cali Finds the ABCs by Terre Redmon

Penny the Parakeet and the Solar Eclipse by Melody L. Schaefer

Check out these Children's Books from

Amazing Things Press

Potty Party! by Andrea Leigh Plummer

The Alphabet Gang Steals the Zs by T.R. Henson

But What If I Miss You? by Stacey Mollus

Ellie Fant by Kathy Blair

Finding Hope by Andrea Leigh Plummer & Steve Sewell

Finding Hope Grief Activity Book by Andrea Plummer & Steve Sewell

Strong Daddy by Andrea Leigh Plummer

Greta the Goose by Connie Condron Dow

God's Tiny Angel by T.R. Henson

Dallas Has Puppies by Terre Redmon

Chrissy the Frog by Kathy Blair

Radiant Red: The Worry Wand by Vanessa Culver

The Far Away Princess and the White Buffalo by Thomas Kirschner

MOBugs: Butterflies and Moths by Shelly Cox

Penny the Parakeet's Royal Adventure by Melody L. Schaefer

More Casey's Creatures by Jonn T. Casey

Amazing Things Press

www.amazingthingspress.com

Made in the USA
Monee, IL
05 November 2026

67985532R00017